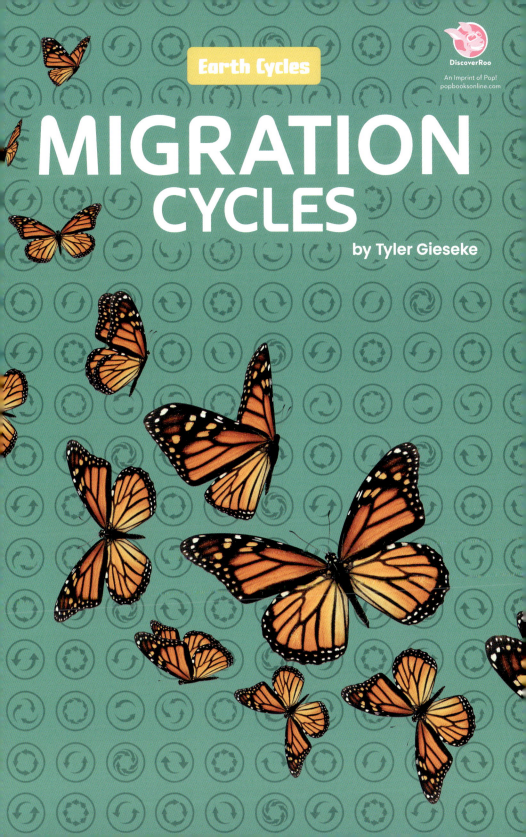

abdobooks.com

Published by Pop!, a division of ABDO, PO Box 398166, Minneapolis, Minnesota 55439. Copyright © 2023 by Abdo Consulting Group, Inc. International copyrights reserved in all countries. No part of this book may be reproduced in any form without written permission from the publisher. DiscoverRoo™ is a trademark and logo of Pop!.

Printed in the United States of America, North Mankato, Minnesota.

052022
092022

THIS BOOK CONTAINS RECYCLED MATERIALS

Cover Photo: Shutterstock Images
Interior Photos: Shutterstock Images
Editor: Elizabeth Andrews
Series Designer: Laura Graphenteen

Library of Congress Control Number: 2021951858
Publisher's Cataloging-in-Publication Data
Names: Gieseke, Tyler, author.
Title: Migration cycles / by Tyler Gieseke.
Description: Minneapolis, Minnesota : Pop, 2023 | Series: Earth cycles | Includes online resources and index
Identifiers: ISBN 9781098242206 (lib. bdg.) | ISBN 9781098242909 (ebook)
Subjects: LCSH: Animal migration--Juvenile literature. | Animal migration--Climatic factors--Juvenile literature. | Animal migration--Endocrine aspects--Juvenile literature. | Earth sciences--Juvenile literature. | Environmental sciences--Juvenile literature.
Classification: DDC 591.52--dc23

Pop open this book and you'll find QR codes loaded with information, so you can learn even more!

Scan this code* and others like it while you read, or visit the website below to make this book pop!

popbooksonline.com/migration

*Scanning QR codes requires a web-enabled smart device with a QR code reader app and a camera.

TABLE OF CONTENTS

CHAPTER 1
There and Back Again 4

CHAPTER 2
Chasing Comfort 10

CHAPTER 3
Home to Breed 16

CHAPTER 4
Hoofing It. 22

Making Connections. 30
Glossary 31
Index. 32
Online Resources 32

CHAPTER 1
THERE AND BACK AGAIN

Most people know that many birds fly south for the winter. They go somewhere warm. Then, they are comfortable. They travel north again when spring comes. Birds migrate every year.

WATCH A VIDEO HERE!

Birds often travel together in flocks when they migrate.

Other animals migrate, too. This means they leave one place, travel somewhere else for a while, and eventually return to the first place. They do this over and over again, usually once per year. This is a migration cycle.

Many types of penguins follow migration cycles.

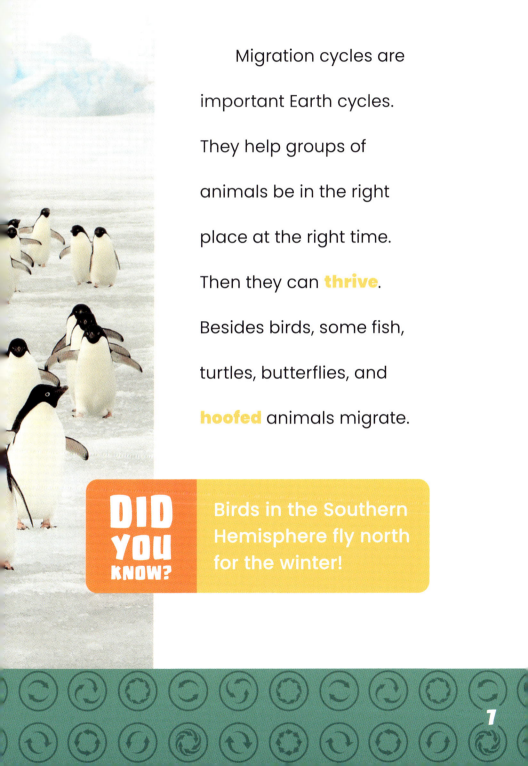

Migration cycles are important Earth cycles. They help groups of animals be in the right place at the right time. Then they can **thrive**. Besides birds, some fish, turtles, butterflies, and **hoofed** animals migrate.

DID YOU KNOW? Birds in the Southern Hemisphere fly north for the winter!

Even tiny butterflies can travel long distances during migration.

There are several reasons animals migrate. Some travel so they are warm and comfortable during winter. Some return to a shared place when it is time to **breed** or give birth. Others do a mixture of both. Migration cycles are everywhere. It is important to understand them.

A MIGRATION CYCLE

A typical migration cycle starts in a **species'** breeding ground. The animals have babies in the summer. They travel south in the fall for warmth. Then, they spend the winter there. This is called overwintering. Finally, the species migrates back to its breeding ground in the spring. The cycle starts again!

CHAPTER 2
CHASING COMFORT

A common reason for migration is to escape the cold or travel somewhere more comfortable. Canada geese and monarch butterflies migrate for this reason.

LEARN MORE HERE!

Canada geese are common in North America.

Monarch butterflies cover the trees in Mexico.

Canada geese **breed** in the northern parts of North America during the warmer months. They fly south in the fall. They spend the winter in the southern United States or northern Mexico. They can stay warm there. When the weather heats up in the spring, they head back up north. The round trip could add up to 3,000 miles (4,800km) or more!

Even though they are much smaller, monarch butterflies make a similar trip. These bugs fly from the United States to spend the winter in central Mexico. Some will cross the Gulf of Mexico to get there. Once they arrive, large numbers of them cover tree trunks. They move very little so they can **conserve** energy until spring.

DID YOU KNOW? Migrating monarchs can cover as many as 80 miles (130km) in one day.

A group of penguins on land is a waddle.

On the trip home, each butterfly only lives for a few weeks at a time. They lay new eggs along the way. Then, the

newborn butterflies continue the journey! The butterflies that complete the trip north might be the great-grandchildren of the bugs that spent the winter in Mexico.

WADDLE OF THE PENGUINS

Emperor penguins in Antarctica migrate 100 miles (160km) inland each autumn to lay eggs. After a male and female breed, the female returns to the ocean for food. For three months, the male remains at the breeding ground and keeps the egg warm under his feathers. He doesn't eat! The female usually returns after the egg hatches. Then, the male waddles to the ocean to feed.

CHAPTER 3
HOME TO BREED

Animal groups also migrate so they can **breed** and raise babies. Their children might need to grow up under certain **conditions**. Or, the animals might head home to breed simply out of habit.

EXPLORE LINKS HERE!

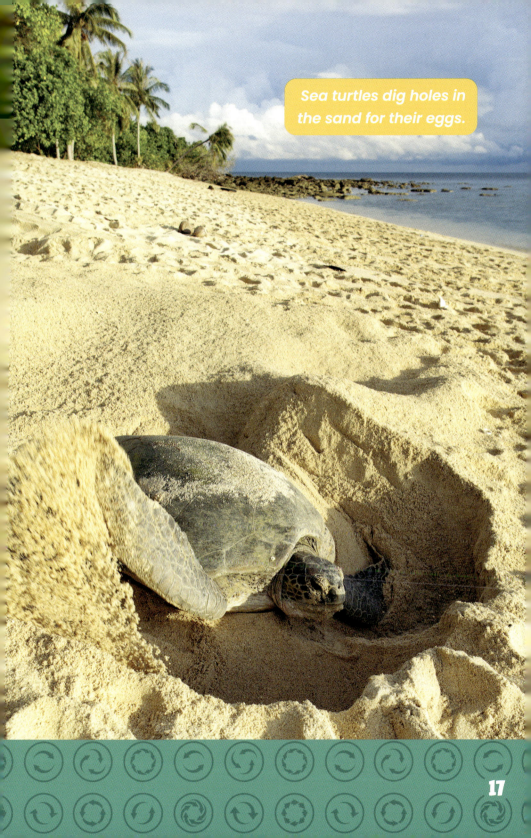
Sea turtles dig holes in the sand for their eggs.

Salmon migration can be beautiful.

Salmon make up one **species** that migrates to breed. Baby salmon hatch in the fresh water of rivers or streams. As they get older, they travel to the salt water of the sea.

When it is time to breed, they go to the stream or river where they were born and travel back up it. This requires a lot of strength from the salmon. They breed and lay eggs in the fresh river water. After breeding, some kinds of adult salmon die. Other types return to the sea.

Sea turtles make some of the longest migrations of all animals. They swim very far to find jellyfish to eat. Still, they can find their way back to the beach where they were born. They will breed there. Scientists are not sure why the turtles go back to the same beach to have babies. It may just be a habit!

One group of scientists put a tracker on a sea turtle in 1996. The scientists let the turtle go free on the west coast

DID YOU KNOW? Scientists think sea turtles find their way home by reading Earth's magnetic field and the ocean water around them.

Sea turtles can ride ocean currents to save energy.

of North America. The tracker showed that the turtle crossed the entire Pacific Ocean. After a trip of more than 5,800 miles (9,300km), the turtle ended up near Japan.

CHAPTER 4
HOOFING IT

Some **hoofed** animals migrate in very large numbers, often to find the best places for food. In East Africa, millions of wildebeest migrate in a circle around Serengeti National Park. In Alaska, hundreds of thousands of caribou

COMPLETE AN ACTIVITY HERE!

travel south in the winter. They return to northern **breeding** grounds in the spring and summer.

Wildebeest and zebras fill the Serengeti plains.

Crocodiles hunt migrating wildebeest.

The movements of the Serengeti wildebeest are called the Great Migration. To find the best **grazing** plains, they

move north during the first half of the year. They spend the summer near the border of Tanzania and Kenya. In the second half of the year, they slowly return to the south.

The Great Migration is extremely dangerous. The trip is long and difficult. There are many **predators**, including crocodiles and cheetahs. Hundreds of thousands of wildebeest die during migration each year.

On the other side of the world, the caribou of Alaska face fewer predators. But, like the wildebeest, they migrate throughout the year. Scientists estimate that 250,000 Alaskan caribou migrate 2,000 miles (3,200km) every year. In the winter, they head south to eat **lichen**.

Caribou often travel in large groups.

In the warmer months, they go to the far north and breed.

Migration cycles are important Earth cycles. They help animals **thrive** and grow wherever they live on the planet. Migration cycles make up just one of the many cycles that shape our world.

 **DID YOU KNOW?** The Western Arctic Herd of caribou lives within an area about the size of California.

There are migration cycles in all parts of the world. Some animals, like the saiga antelope in Asia, travel relatively short distances. Others, like sea turtles and southern elephant seals, travel quite far. Do any of these migration paths surprise you?

MAKING CONNECTIONS

TEXT-TO-SELF

Which was your favorite animal to learn about in this book? Why?

TEXT-TO-TEXT

Have you read other books about migration? What new information did you learn in this Earth Cycles book?

TEXT-TO-WORLD

Do some humans migrate? Try and think of some examples. Why do people migrate?

GLOSSARY

breed — to make new animals of the same kind.

conditions — features of an area, like warmth or saltiness.

conserve — to use just a little bit, so a stored amount does not run out.

graze — to eat grass in an area.

hoofed — having a hard covering on the ends of the feet.

lichen — a fungus that grows on plants or rocks.

predator — an animal that hunts other animals for food.

species — a type of living thing.

thrive — to grow and live with good health.

INDEX

Alaska, 22, 26

breeding, 8–9, 12, 14–15, 16, 19–20, 23, 27

Great Migration, 22, 24–25

Kenya, 25

Mexico, 12–13, 15

North America, 12–13, 21

Pacific Ocean, 21

Tanzania, 25

warmth, 4, 8–9, 12, 27

ONLINE RESOURCES
popbooksonline.com

Scan this code* and others like it while you read, or visit the website below to make this book pop!

popbooksonline.com/migration

*Scanning QR codes requires a web-enabled smart device with a QR code reader app and a camera.